PLOTTING A COURSE

THE MEMOIRS OF
J. U. M. SMITH

PLOTTING A COURSE

THE MEMOIRS OF
J. U. M. SMITH

ISBN 9781917109123

Published 2024 by Tricorn Books
www.tricornbooks.co.uk
Treadgolds 1 Bishop Street Portsea
Portsmouth UK

PLOTTING A COURSE

Contents

Chapter One
A WARTIME CHILDHOOD

Brixham 1939–41

"That's a German plane!" said my grandmother.

"Nonsense," replied my grandfather, "it's a lorry coming down the hill."

But Granny was right; it *was* a German plane! Luckily it passed over and did not return.

It was the summer of 1940 and I was seven years old. My mother, my elder brother Christopher and I were sitting at the dining table in my grandparents' house in Brixham, south Devon, having our midday meal. Granny at the head of the table, presiding, and Popper – we always called grandfather Popper – sitting opposite me with his bottle of 'porter' (beer), in defiance of his wife who disapproved of alcohol.

After the fall of France, Brixham had come within the range of German bombers based on the Cherbourg peninsula. We were on the front line. This was ironic because the only reason we were in Devon in the first place was to escape the bombing in London where we had been living!

The raids were frequent. They must have been frightening for the grown-ups, but we children were more interested in the aircraft themselves. We had an aircraft recognition book which gave details of all the German aircraft and the three silhouettes of each – plan, side and front. It turned out that the raiders were mostly Messerschmitt 109Fs.

We had some narrow escapes. Once we got caught on the beach by the open-air bathing pool. We could see the enemy aircraft across the bay bombing Torquay. Then it turned towards us.

"Quick," said my mother, "come and shelter in this rain shelter."

And there we sat – sitting ducks because the rain shelter had no sides. It was completely open to the enemy's gunfire. The plane was continuing to turn towards us when we suddenly heard an anti-aircraft gun quite nearby open up. The pilot obviously thought better of coming for us and turned back to the open sea and home.

On another occasion I was with my father on the beach when we heard the sound of aircraft engines. He knew exactly what to do. He had served in the trenches in the First World War.

"Quickly, get down flat on the rocks," he ordered as the planes roared harmlessly overhead. Early lessons on how to behave under fire.

At that time there was an old coal bunkering ship called the *London City* permanently moored alongside the Breakwater. I believe this was one of the reasons for the frequent raids on Brixham. The Germans obviously thought she was a strategic asset and were determined to sink her. This was a mistake on their part because most ships were by then oil powered and used instead the oil jetty at the end of the Breakwater which was never attacked. The *London City* was sunk three times in all, each time being refloated. The harbour was so shallow that she would settle on the seabed with the uppers still visible. At low tide she would appear to be still floating which must have confused the enemy pilots who thought they had already sunk her. There was a rumour that the *London City* was originally a German vessel ceded to Britain as part of reparations after the First World War, and this is why the Germans were so keen to sink her. But this may have been only a rumour.

These tip and run raids came suddenly and without warning. The aircraft came in low, dropped their bombs, fired their machine guns and departed out to sea, hugging the waves to avoid our guns. They did not do a lot of damage, although they destroyed some houses and hit a church in Torquay, all of which was hushed up at the time. After Hitler invaded Russia in 1941, the raids became fewer and eventually stopped.

***

I was born on the 28th of April 1933 in Hampstead Garden Suburb, in northwest London, where the family lived until the outbreak of war. My father was a teacher in an inner London school. In 1939 his school was evacuated to Winsford, a tiny Somerset village, to escape the forecast bombing of London. Winsford was too small to accommodate the rest of the family which is why we ended up with my maternal grandparents in Devon when war broke out.

I don't remember much of life before the war. One of my first memories is of Mr Chamberlain's declaration of war on 3rd September 1939 – "For it is evil things that we shall be fighting against, brute force, bad faith …" The whole family was sitting around the radio in an upstairs sitting room at my grandparents' house. But do I really remember this? Or is it that I have heard the recording so many times since?

Of course I didn't then know what war meant and asked my big brother Christopher.

"Oh," he declared authoritatively, "you will have to drive a tank, you know."

I did not like the sound of this; I was quite worried until my mother said, "No, children do not have to go to war. War is for grown-ups." What a relief.

Later that day, the air raid siren went off. The whole family, grandparents included, trooped down to the coal cellar which acted as our air raid shelter – I can still remember the smell of coal – and sat obediently until the all-clear sounded. After a while sitting still, my father and Christopher were allowed, very daringly, to venture forth to see what was happening. I was forbidden to accompany them, something I bitterly resented. Of course nothing was happening. The Germans were far too preoccupied with the invasion of Poland to bother sending planes over to a remote seaside town in Devon.

Whether I remember the Declaration or not, I *do* remember the first days of the war and the preparations that we made. We were all issued with identity cards and gas masks. I don't remember ever having to wear my mask, which was a nasty uncomfortable rubber thing, but certainly I had to take it with me in its case when going to school. We were also issued with a sand bucket and a long shovel for putting out fire. We had to stick netting or strips of tape on all the windows to protect against shattered glass. The blackout was imposed immediately, and a warden cycled around shouting, "Put that light out!"

Then there was the rationing – of just about everything from clothes to petrol and food. Ration books were issued to the grown-ups and they organised the distribution within the family. We each were given little dishes for our portion of butter and sugar and so on. The portion had to last a week so there was the question whether to have it all at once or make it last. Luckily we had a great aunt who lived on a farm at Churston, not far away, and our meagre rations were sometimes supplemented by a surprise visit bringing some additional food.

Every day the grown-ups listened to the BBC news to hear how the war was progressing but we children were far more interested in going down to the beach. I do remember the news coming in that the German pocket battleship *Admiral Graf Spee* had scuttled herself off the River Plate in December of 1939, having been chased into Montevideo harbour by the Royal Navy. My father was delighted.

"The war is nearly won!" he said. An example of his dry sense of humour!

Looking back now, our stay in Devon seems like the best of times. We children were free to roam without adults telling us what to do. We could explore the coast and the surrounding countryside as we liked. In those days the sea was full of fish. In the summer, mackerel drove shoals of 'brit' (whitebait) inshore while porpoises out in the bay in their turn pursued the mackerel. The mackerel were so plentiful that you could catch them in a net. There were prawns to be caught at low tide and even crabs and lobsters if you were clever and knew where to look. We went mushrooming on Churston golf course in the early morning when no one was about, and once caught an adder on Berry Head and brought it home in a sack, much to the consternation of the grown-ups. We put a forked stick behind his head, picked him up by the tail and dropped him into the sack.

Grandfather, John Upham, was an imposing figure. He had snow-white hair, an equally white neatly clipped moustache and a twinkle in his blue eyes. Then retired, he had been the senior partner in the family shipbuilding firm *J W and A Upham* which had built the famous Brixham sailing trawlers in the nineteenth and early twentieth centuries and was now, in wartime, busy churning out small naval craft including an armed motor launch. He always wore a dark three-piece suit with a pocket watch on a chain in his breast pocket. He never took much notice of me except occasionally to lean across the dining table, point to me and say, "That boy needs a haircut."

Granny on the other hand took a great interest in me and in her large family. She was the daughter of the local butcher and had married well. She had had six children – five girls and one boy – and now had countless grandchildren and great-grandchildren which she was always muddling up.

After attending a couple of dame schools in Brixham, I was sent to the Edward the Sixth Grammar School in Totnes, which was then about halfway up Fore Street on the left-hand side. It was quite a long journey: walk to Brixham Cross, bus to Paignton then another bus to Totnes. I don't remember much about the school and I don't think I learnt much there. The schoolmistress was an ignoramus. For instance, she thought that William of Orange had landed at Torquay. I tried to correct her but she was not having it. I knew I was right. After all, I was familiar with William's statue on the Brixham Quay inscribed with his famous pledge that he would maintain English Freedom. I had also seen the stone slate on which he was said to have stepped as he came ashore in Brixham harbour in 1688.

By the summer of 1941 the war danger had receded somewhat and it was judged safe to return to London. The Blitz was over and Hitler had turned his attention to the East with the invasion of Russia. The threat of invasion and the large-scale destruction of London did not seem so real. My father's school decided to return and so the family prepared to return to our home in Hampstead Garden Suburb.

While our parents were in London getting the house, which had been left empty, in order, Christopher and I stayed in Devon boarding at the school. At Christmas we travelled back to London, taking the train from Totnes station. We were met at Paddington by our parents and conducted home on the underground and bus to Golders Green. Thus ended our two-year sojourn in the West Country.

Hampstead Garden Suburb 1941–1945

Although the Blitz was over, German bombers still came over. We had a Morrison shelter, named after Herbert Morrison the then Minister of Supply. This was a massive metal table with thick legs and netting sides set up in the dining room, strong enough to protect us against falling masonry, even perhaps a direct hit. The idea was that it could act as both a dining table and a shelter. Most people had the Anderson type of shelter which was a dugout shelter half buried in the garden. This would have afforded better protection than our Morrison. The Morrison was presumably designed for those who did not have a garden. I don't know why we did not have an Anderson. Perhaps it was because we had been away for the first two years of the war, or perhaps it was because my father did not want to spoil his neat garden. My mother and us two boys slept in the Morrison while Dad slept under the stairs. Mum made sure she had all the important papers with her when she went to bed at night. We didn't do this for long. We soon reverted to our own comfortable beds upstairs and got rid of the ugly Morrison encumbrance.

The air raid sirens went off fairly frequently at this time. Most bombs fell a fair distance away but we did have some quite close ones. Luckily, we were never hit. I do remember at least on one occasion hearing the whistle as they fell. And waiting with bated breath for the inevitable explosion. On the morning after a raid, Christopher and I would take to our bicycles and search for shrapnel – fragments of exploding AA shells. We acquired quite a collection of these which we kept in a large wooden box. They were sharp-edged pieces of metal of varying sizes with rifling on their surfaces. The larger pieces were the more valued ones. One day we found the tailpiece of a bomb sticking out of the ground. Christopher went to investigate. I tried to stop him.

"That's an unexploded bomb. It might go off," I warned him.

But no, he went ahead and pulled it out of the ground. Luckily it was only the tailpiece without the warhead. Phew! Another piece of luck!

My first school was Lees House, a private prep school on the other side of the Suburb within easy cycling distance of home. There I was introduced to, among other things, Latin and algebra. Then after a year I entered Haberdashers' Aske's boys' day school, at Cricklewood. Haberdashers' was an independent grant-aided school founded by the Worshipful Company of Haberdashers, an ancient City Guild. The Guild had several schools; one was a girls' school at Hatcham in South London with which, later on, we had joint social events – very exciting but also somewhat terrifying. I started in the first form and stayed on for eight years, ending up in the senior sixth form before moving on to college as I shall explain in the next chapter. I journeyed to school each day, either taking the trolley bus down the Finchley Road, or by bike, depending on weather and my inclination.

Hitler's V weapons

In the summer of 1944, the war seemed to be going well. D-Day had been a success, the Russians were advancing on the Eastern Front, and the Americans were winning the war in the Pacific. But Hitler had one nasty trick up his sleeve: his V weapons – his *Vergeltungswaffe* or reprisal weapons. They were marvels of technology in the predigital age, but deadly.

There were two types: the V1 flying bomb, affectionately known as the doodlebug, and later the V2 ballistic rocket. The V1 was a small pilotless aircraft with a large explosive charge in its nose, launched from a fixed ramp pointed directly at London. The thing about the V1 was that it was quite different from a manned bomber. In the case of the latter, if you heard it you knew you were in trouble. But if you could hear the characteristic sound of the V1's pulse jet engine, a bit like a motorbike engine, coming, you knew you were safe. It was when you *didn't hear* it that you had to worry. The V1 was programmed to cut its engine over the target and glide to earth. So when you heard the engine stop that was when you had to take cover.

On the first few nights, the AA guns put up a tremendous barrage of counterfire. But then suddenly one night there was absolute silence, no gunfire at all. We could not work out what was going on. The explanation was that the authorities had realised that there was no point at all in attempting to down a flying bomb over a built-up area. After all, that was what they were designed for – to come down on a built-up area. The guns had been moved south of the capital in an attempt to destroy the V1s before they could reach their target. Fighters were also stationed south to intercept them and, in some cases, turn them back.

We were fortunate to live in north London. Most V1s landed to the south or in the centre. So we got off relatively lightly. But we did have one or two close shaves. One came over when I was at school. The school had an air raid shelter built on the playing fields just behind the school building. Senior boys were posted on top of the science block as lookouts. When the siren went we were supposed to file out of our classrooms in an orderly fashion and make our way outside to the air raid shelter. I vividly recall one occasion.

We were in the open on our way to the shelter when we heard the V1 coming overhead. Then its engine cut out. Silence. Some boys began to run. I flattened myself on the ground and waited. *It's coming*, I thought. *It's going to land somewhere near*. Then there was an almighty bang. The bomb had landed next door. Missed us! Another lucky escape.

As a result of this event and other similar ones, my mother decided to stop us going to school on cloudy days because it was on those that the Germans were most likely to send the V1s over as it gave them the best chance of avoiding our artillery and the fighter defences. My mother's philosophy was if we were going to go, we would go together as a family. Consequently, we boys looked forward to bad weather; it meant we could have a day off school.

By the autumn of 1944, the Allies' advance on the continent had overrun the launching sites in France and Belgium and the V1s stopped coming. But we were not out of the wood. The Germans now launched their V2s on us. The V2 was a liquid-fuelled rocket that rose into the stratosphere and then fell from a great height onto its target. It carried a massive explosive charge that did extensive damage. There was no known means of defence, except to hit the launch sites, but these were hidden amongst trees and moved around.

The peculiarity of the V2 – unlike the V1 it did not have a pet name – was that it arrived at its target before you heard it coming. You heard a loud explosion then a rumbling sound which was it descending at supersonic speed through the atmosphere. We had no warning at all of a V2 strike. It could come anywhere at any time. You just hoped it wouldn't be where you were.

The end of the war

One day, as I was walking up Temple Fortune Hill, a woman poked her head out of an upstairs window.

"Have you heard the news? Brussels has been liberated!" she shouted.

At this point we thought the war was almost over. But then there was no more news. The BBC was silent for months – it was the Battle of the Bulge, the final German push through the Ardennes. As an aside, it is difficult for the modern reader, with all the stories that have been told since, to realise how little we knew of what was going on at the time. My mother didn't like the absence of news.

"Something has gone wrong," she said.

But then in the following summer, victory finally came.

We went up to the West End to join in the celebrations. I can remember walking up Whitehall with the crowds to Trafalgar Square and up The Mall to Buckingham Palace. The whole of the West End was free of traffic and the crowds on foot were immense. We stood in front of the Palace and saw the King and Queen appear on the balcony with Winston Churchill, the hero of the hour. There were immense cheers from the crowd. The King, who had taken no part in the war, was resplendent in the uniform of Admiral of the Fleet, the highest military rank. Churchill, on the other hand, who had directed the whole war effort, was dressed in modest civilian clothes. The BBC came on the loudspeakers and announced that an old friend, a deep depression, was approaching from the Atlantic. This raised a great laugh from the crowd. We hadn't had weather forecasts during the war. When it was time to go home, the crowd dispersed all at once through the narrow entrance to Buckingham Palace Road. People were being pressed together and I felt as though I was getting crushed. There was no crowd control in those days and there could have been a disaster.

Reflections on wartime

As a family we came through the war quite well. My father was too old to fight, although he was in the Home Guard, and we children were too young. We did however lose an older cousin, Peter Hellyer, who was killed on active service with the Fleet Air Arm. His mother, our Aunty Frances, never really recovered. And we had a Brixham acquaintance who went down with the *Rawalpindi*.

During the war there was a sense of community, of all pulling together. We had a single goal: win the war. Now that peace was here, what was the objective? I still today sometimes ask that question.

We were never hungry, although rations were meagre by modern standards. And we never thought we were going to lose the war. The only time I remember the grown-ups being worried was when the German pocket battleships, *Scharnhorst* and *Gneisenau*, made their successful dash up the channel in 1942.

"That should never have happened," opined my mother. But if you read Churchill, you realise how close we came to defeat in the summer of 1940 when invasion was a real possibility and we had left all our armour and equipment behind at Dunkirk.

Chapter Two
EDUCATION

Latin

We had Mousey for Latin. Mousey was not his proper name but we always we called him Mousey. He had probably been a brilliant scholar in his day and knew all the different modes and subtle uses of the gerund. But he was a hopeless teacher and couldn't keep order in the classroom. He was an object of ridicule, poor man. We boys invented a phantom pupil named Bill Sykes. At every roll call someone would answer for Sykes, or else explain his absence. We kept this game up for a whole term before Mousey twigged.

We were coming up to the dreaded School Certificate exam – O-level equivalent – and the syllabus comprised Caesar's Gallic War book 1 and Virgil book 4. So we started reading all about all Gaul being divided into three parts etc. But then over the Christmas holidays one of the parents discovered that the syllabus was actually the other way around, not Caesar 1 and Virgil 4 but Caesar 4 and Virgil 1. Consternation. So we had to start again. But perhaps it was all to the good because it broadened our education. In addition to learning how many parts Gaul was divided into we also learnt how Caesar managed to build a bridge across the Rhine.

I was not learning much from Mousey so my mother paid for some private lessons in the Easter holidays and this was just enough to get me through the exam with a credit which was fortunate because it enabled me to apply to Cambridge

University later on. Latin was a requirement in those days for the ancient universities.

Mathematics

I had always liked maths. When once I was given a detention, the master in charge of detention gave me a set of quadratic equations to solve. I actually enjoyed doing these and asked for some more please. This was not a punishment. It was fun.

When we got onto the differential calculus I was hooked. We had a very good maths teacher. His mantra was 'find the questions you can't answer and answer those'. He taught us to seek the shortest, most economic and most elegant answers.

After taking the School Certificate, we had to choose between science subjects and the arts. I chose the former. The science course normally consisted of physics, chemistry and mathematics, but I decided to do geography instead of chemistry. Although I did not realise it at the time, this decision steered my academic career towards a mathematics degree because for natural science, the degree that most boys did, you needed chemistry. However, this choice did have some advantages. When I got to college, I found I did not have to spend afternoons messing about in the laboratory. I could spend the time more enjoyably on the river, rowing.

The Air Training Corps

One day I was on my way to the school playing fields at Mill Hill, pushing my bicycle along a footpath on high ground overlooking Hendon aerodrome, not thinking of anything in particular, when *whoosh!* an aircraft swept overhead on its final approach to landing. It could not have been more than a few feet above me. I followed it with my eyes all the way down to the runway. I was fascinated. *How marvellous it must be to glide through the air like that,* I thought. Thus began my lifelong interest in aviation. I did not know then that in a few years' time I would be in a similar plane coming in to land in exactly the same place. (It was probably an Avro Anson.)

In those days, most schools had a cadet corps and Haberdashers' was no exception. Most boys joined. I joined the air force section – the Air Training Corps (ATC). We met every Friday afternoon. We wore proper RAF uniforms and badges. So on Thursday night I had to blanco my belt and polish my cap badge. The school had its own firing range where we could practise firing .22 calibre rifles. Pull the butt into your shoulder, hold the rifle steady and gently squeeze the rigger. But I was more interested in what we were taught about the principles of flight than in the military drills. I started reading articles in aviation magazines and found out how lift was produced (via accelerated airflow over the upper surface of the wing).

The best part of corps was field day which happened once a term. Whereas our army section friends spent the day tramping across or crawling over damp muddy fields, we got to go flying. Often we went to Fairoaks Airfield where they had Tiger Moths open cockpit biplanes and were taken up for a flip. One time I remember well the pilot asked if I would like to try aerobatics. Of course I said yes. After gaining sufficient height and circling to make sure the sky was clear he put the plane into a steep dive to pick up speed for a loop. Then just as he was pulling back to start the loop, my shoulder straps slipped off my shoulders. Help! I shall fall out when we get to the top of the loop! But the pilot kept positive *g* all the way around so I did not fall out. Another lucky escape! In retrospect, I would probably have been saved by my lap straps anyway.

Occasionally on a Saturday we were allowed to go to Hendon where we would be taken up in an Avro Anson twin-engine communications aircraft. These trips were supposed to be navigation exercises, but since we were not given maps they were actually pure joy rides. So here I was in the same type of plane coming in over the same footpath that I had stood on years before when I first started getting interested in the marvels of flight. In a funny way, I had now come full circle.

The French Connection

Our next-door neighbours were two middle-aged single ladies, Miss Betts and Mademoiselle. We never knew Mademoiselle's real name. She was French and would sometimes engage us in conversation though the hedge that divided our two gardens. Occasionally she would switch to French, possibly for my benefit, whereupon my mother would throw her hands up in horror and flee into the house.

We were a strictly monoglot family. To remedy this my parents arranged for me to have a French pen pal, Jacques. To start with we merely exchanged letters. Later we exchanged visits. I remember the letters coming, written on strange, squared paper mostly in French. We could not understand why Jacques was always telling us that he had just been to relieve himself. It was some time before we realised that *piscine* actually meant public swimming pool.

Jacques made the first move by coming to stay with us. Looking back, it must have been quite brave of him at age thirteen to come all alone across the channel to a strange country to meet strangers. I remember going down to Victoria Station to meet him. We waited at the platform barrier for the boat train to arrive. The train came in. All the passengers got off and filed past us through the barrier.

No Jacques. Then at the very furthest end of the train a very small figure appeared, weighted down by an enormous backpack. This was Jacques. He turned out to be a rather annoying individual who kept on asking, "When do we eat?"

We took Jacques to all the usual tourist spots – Buckingham Palace, the Tower, etc. We even visited the House of Commons and saw Winston Churchill speaking. Then it was my turn to make the return visit. His family lived in the little village of Saint-Cyr-sur-Loire just outside Tours. Jacques's parents were both teachers. They ran the village school and lived above the school. His father was a jovial individual with a twinkle in his eye. He claimed to be a sportsman although he played no games.

"I have driven many miles in my motorcar," he said, "isn't that a sport?"

My first impressions of France were what a strange place it was. We arrived at Dieppe off the boat and went to the station to catch the train to Paris. But where was the station platform? There wasn't one. You had to climb from the ground up a ladder to enter the carriage.

Jacques's father met us in Paris. To fill in time before the train to Tours, he took us to a small pavement café opposite the Opera. A street trader approached our table and tried to interest Jacques's father in some leather goods. Jacques's father inspected a wallet.

"Yes," he said, "it is certainly a fine wallet, but I am only a poor schoolteacher and I can't afford the price."

I expected that to be the end of the conversation. But no. Jacques's father kept on praising the wallet and pleading poverty until the trader reduced the price. This process went on for some time until Jacques's father finally agreed to buy the wallet. My first lesson in haggling.

That first visit was in 1947, only two years after the end

of the war. France was still recovering. The bridge over the Loire was still a temporary structure; the original having been blown up by the retreating Germans. When I went back two years later there was a new bridge in place. Enroute, we stayed in Paris for a couple of days. I remember going to see Napoleon's tomb in *Les Invalides*. Jacques's mother was not impressed. He killed many people, was her comment.

United States

Now a stupendous piece of good fortune came my way. I got to go on a visit to America, all expenses paid. A philanthropic youth organisation in Nashville, Tennessee, called Youth Incorporated, had decided to augment their summer camp programme by inviting a group of youngsters from across Europe to visit and take part. I was put forward by the school and was lucky enough to get the gig. The press got hold of the story and I had my fifteen minutes of fame when the *Evening Standard* published a piece entitled 'Lucky dog they say in the sixth'.

We flew across the Atlantic in a Lockheed Constellation, a four-engine propeller plane, one of the first to have a pressurised cabin. Remember this was 1950, early days in civil aviation. Heathrow had only just opened and consisted merely of a few temporary huts on the north side alongside the Bath Road. The Constellation did not have the range to do a non-stop London–New York so we made stops at Gander, Newfoundland and Boston before arriving at New York. The next day we flew onwards to our destination, Nashville.

I was invited to stay at one of the sponsor's houses. I was met by a large smiling *African American* who held out his hand in what I thought was greeting. I naturally took his hand and shook it. Big mistake. This was the South and segregation was still very much in place. He was not greeting me. He was one of their servants and was merely offering to take my suitcase. What a gaffe. Goodness knows what my hosts thought of it.

We returned from New York on a Boeing Stratocruiser, a double-decked plane with a spiral staircase leading to a comfortable downstairs lounge. State of the art at the time.

Cambridge

After taking my Higher School Certificate exams, the equivalent of A-levels, it was time to think about university. I had always set my sights on Cambridge. I chose King's College because it included physics in its entrance exam. I went up for the interview, took the scholarship exam and won an Open Exhibition. And so it was that in the autumn of 1952 I found myself on a train going up to Cambridge and starting a new chapter of my life.

Cambridge at that time was largely a male, public school place. There were I think only three women's colleges. King's was full of old Etonians. There was quite a division between those undergraduates who had already done their national service and those like me who had come straight from school. The former were more mature and tended to run the societies and other activities.

In my first year I lived at the Peas Hill Hostel, just across King's Parade from college, next to the Arts Theatre. It was an old building with sloping floors and rather primitive washing facilities. The partition walls were thin. I could hear Tam Dalyell next door rehearsing his speeches for that evening's debate at the Union. Tam was a baronet with a country seat on high ground just south of Edinburgh, with views over the Firth of Forth and peacocks on the lawn. The only time he deigned to speak to me was when he was up for President of the Union and wanted my vote. (He failed to get elected.) But then why should he take any notice of me, a mere commoner from the suburbs? Dalyell went on to become an MP, becoming famous for asking the West Lothian Question, which still hasn't been satisfactorily answered, and for becoming mixed up in the Belgrano affair.

As an Exhibitioner I was allowed to spend my next two years in college itself. My first college rooms were in the 'drain', an old building, long demolished, across the King's Lane from the main college buildings and only accessible by a tunnel. On inspecting my new lodgings, I found to my surprise an upright piano. I had played a bit before – so I started practising my scales again and persuaded a friend to play duets. However, when I returned after the Christmas break I found the piano had disappeared. Apparently it was only a hired instrument and it had been repossessed during the holidays. However, this was not the end of my musical endeavours. I had another friend, very musical, and we found a practice room that we could hire. He arranged some of Bach's four-part fugues for two pianos and we had great fun with these. We took two voices each which obviously simplified the sight reading, but needed attention to the synchronisation.

I took all my meals in college. We were fed three times a day. The evening meal was quite an affair. We assembled in Hall at 7.30 pm. Then the dons entered and processed to the High Table. When everyone was seated, one of the senior scholars would say grace – in Latin of course. The Steward would sound the gong whereupon a team of waiters would appear from the kitchen bearing soup, and the meal would commence. On special days, such as Founders Day, we had a feast. A loving cup would be passed around the table from person to person – most unhygienic. It was supposed to still contain some of the original wine served when the college was founded 400 years ago.

The food in Hall was surprisingly good, considering that we were still living under rationing. Attendance was mandatory. You had to get permission if you wanted to eat out. Indeed, one got the impression that eating the requisite number of dinners was as important, if not more so, than passing examinations. A most civilised approach to acquiring a higher education!

Lectures were held in the Old Schools building not far from college. There were two types of lecturer; those that could explain the subject succinctly, and those that couldn't. Hermann Bondi, the famous astrophysicist, was of the first type. He strode up and down the lecture hall among the seated rows of us students, and explained in very simple terms difficult concepts such as a vector field etc. He wrote hardly anything on the blackboard. But made it all crystal clear. Derek Taunt, on the other hand, scribbled madly on the board – with his back to us. We had difficulty keeping up with him and writing it all down. In fact, in spite of all the material he was giving us, he didn't finish his course of lectures within term and we had to stay on a few extra days. I sometimes wonder why he did not simply give us written notes. It would have made life simpler and saved a lot of time. Perhaps he feared we would not attend his lectures if we could get hold of the material another way!

My tutor in my first year was Philip Hall, the distinguished algebraist. Every week I would visit him in his extensive suite of rooms on the first floor over the south gate of the college. The rooms had previously belonged to Lord Keynes and had full-length paintings by Duncan Grant on the walls. Philip and I became friends and we would sometimes go to the theatre together along with some other undergraduates that he invited.

One of the advantages of doing mathematics, as I mentioned before, was that afternoons were free; lectures and tutorials were strictly morning affairs. So I joined the College Rowing Club and spent most afternoons on the river. There were races at the end of each term. In the autumn term there was a long-distance race, the Fairburn, and in the Lent and summer terms, there were 'bumps', the latter called 'May' bumps although they took place in June! Due to the narrowness of the river, it was impossible to race side by side so instead boats were spaced out at intervals on the riverbank and the idea was to catch the boat in front before the boat behind caught you. So when the signal was given you rowed like hell. I remember on one occasion the cox of the boat in front got his lines crossed and when the signal sounded his boat sped straight up the bank. That was an easy 'bump' for us. If you got a bump on each day of the races you were awarded your oar to keep as a trophy. I got mine in my first year, and still have it.

In the long vacations, Christopher and I travelled. One year we cycled to Tours to visit Jacques. He had a moped and could easily outpace us on our pushbikes. On the return journey we stopped off at a café for a break. On the menu was *gateau sec*. That sounded good − a nice piece of cake. And it was cheap. Imagine our disappointment when a plate of dry biscuits arrived.

I managed to get a respectable 2.1 degree but this was not good enough in those days to enable one to stay on for a higher degree and do research which was something I would have liked to have done. In retrospect, it was just as well as I don't think I would have prospered in the Groves of Academe.

I now faced national service. I could defer it no longer.

Chapter Three
CANADA

I had a fairly good idea of what I wanted to do, namely to fly. So I applied to the RAF and was summoned to RAF Hornchurch for aircrew selection. There I faced a series of tests, both physical and mental, probably developed during the war, to test my aptitude for flying duties. One I remember involved a peg board. You had to rearrange all the pegs on the board as quickly as you could. This was obviously designed to test manual dexterity. I finished this test well ahead of the others. How? Simple. I used both hands instead of just one!

The result of my visit to Hornchurch was that I was selected for navigator training. This was somewhat of a disappointment as I had hoped to be a pilot. However at least it was flying. I am sure I could have succeeded as a pilot, but the RAF in their wisdom thought that with my mathematical background I would be more use to them navigating which in those days was a calculation-intensive business.

The military services are very good at discovering and exploiting people's strengths. For those with no special qualifications or obvious usefulness, national service was a lottery. I had one friend who spent his two years in the cookhouse, peeling potatoes while another did his in the Navy and got a trip to the Far East. Some saw action. One told me he was posted to Malaya at the time of the insurgency. He had a gun and had to threaten to shoot a deserter who was running away from camp. Most would say they benefited from their time in the 'mob', although perhaps two years was too long. But I digress.

After the summer holidays spent idling in Devon, I received my call-up papers, a travel warrant and orders to report to RAF Cardington in Bedfordshire, which was then the RAF induction centre. We new recruits assembled in the shadow of two huge hangars built in the thirties for the R100 and R101 airships. We were kitted out with our uniforms, and given our identity cards and our service number – mine was 2780307. They say you never forget your service number and I think that's true. There was not much to do while the system decided where to send us. Meantime I whiled away the time lying on my bed reading Tolstoy's *War and Peace*.

Basic training: Kirton-in-Lindsay, Lincolnshire

When they discovered that I had been 'Pre-selected' for aircrew training they posted me to Kirton-in-Lindsay (now simply Kirton Lindsay) in Lincolnshire, for ten weeks of basic training – square bashing. In fact, I had done most of it already at school. There was quite an emphasis on physical fitness. This was tested by the number of sit-ups you could do. It was surprising how many more you could do after ten weeks at Kirton-in-Lindsay.

What was new was the weapons. We spent time on the firing range and learnt how to fire the Bren machine gun and the smaller Sten gun, a hand-held sub-machine gun fired from the hip. It was important, we were told, not to hold your hand over the muzzle of the Sten if you wanted to keep all your fingers. This was sensible advice obviously. What was not so sensible was the advice about what to do in the event of a nuclear attack. Face away from the flash, we were told, if you didn't want to be blinded. Yes but how did you do that when you didn't know when and from which direction the flash was coming?

Kirton-in-Lindsay was an operational station and had an

airfield, but the closest we got to flying, which is what we all wanted to do, was guarding the hangars at night. This involved staying awake all night among the parked aircraft in case an intruder attempted to enter. But staying awake proved almost impossible after an energetic day of strenuous physical activity. We soon became sleepy. But the aircraft were there with their inviting pilots' seats. They were open. So why not climb in and make oneself comfortable? But we were told that every now and then an officer would come around just to make sure we were still awake. This never happened to my knowledge. Nevertheless, it was worrying and prevented us getting a good night's rest. I don't know what the purpose of this exercise was, since we were on a secure base patrolled by military police, unless it was simply to prove that we could stay awake indefinitely.

Oh, I nearly forgot to mention the main thing we were taught at Kirton-in-Lindsay, or rather forced to learn: how to polish our boots, how to get the toecaps so smooth and shiny that you could see your face in them. This required lots of shoe polish, hours spent rubbing it in, in small circles and adding spit as necessary.

Canada

The next part of our training was to take place in Canada. You might have thought we would travel there by air. But no, we went by surface transport. The authorities move in mysterious ways.

We sailed from Liverpool on the *Empress of Canada*. The crossing took five or six days and we luxuriated in first class, dining in black tie and our new uniforms as officer cadets. We felt ourselves no end of swells. We were bound for Winnipeg in Manitoba, where the Air Navigation School was situated. We landed at Montreal after a long sail up the St Lawrence

River and went from there by rail to London Ontario where we stayed for a week to familiarise ourselves with things Canadian. From there on to Winnipeg by train, a journey that took two days, winding slowly across the Canadian Shield, stopping occasionally at request stops where people living in isolated spots got on or off the train.

The first thing we had to do on arrival was to attend a funeral. There had just been a fatal accident that had killed three navigator trainees and their two pilots. This did not seem a very auspicious start to our stay in Winnipeg. Two training aircraft – Beechcraft Expeditors, used for initial training – had collided in mid-air on the landing approach and crashed, killing all on board. It was speculated that the relative position of the two planes in the air was such that neither pilot could see the other. Each aircraft was in the blind spot of the other. This does not explain of course why the control tower did not spot the danger and warn the pilots in time. The funeral took place on the base with full military honours.

This was not the only fatal accident that occurred during our stay in Winnipeg. There was an open day put on for the public. A Fairey Swordfish biplane did a low pass over the crowd. You could clearly see the two crew in their open cockpit. The observer in the rear seat waved to the crowd as he passed. It was the last thing he did. At the end of the runway the pilot banked and pulled a tight left-hand turn. It was a hot summer day. The air was thin. The wings lost traction. The plane stalled and went into a steep dive. The pilot had insufficient height and could not recover. The plane disappeared from view behind some trees and then we saw the ominous plume of black smoke that told us they had not made it. After an interval the show went on.

Of course we knew that flying was a dangerous business

and we were reminded of that every day on entering the ground school building. On the arch above the entrance was inscribed the following.

"Aviation is not inherently dangerous, but to an even greater degree than the sea, it is terribly unforgiving of any careless, incapacity or neglect." (words of Captain A. G. Lamplugh, a famous aviator in the thirties)

But, heck, we were young and took no notice of the risks.

The twelve-month navigation course was in two parts. The first, initial training, was taken by everyone. Then there was a choice – you could opt for airborne interception or for long-range navigation. I opted for the latter.

In initial training we were taught the basics – how to read a compass and correct for magnetic variation and local deviation, how to take a bearing using the radio, how to measure distance on a nautical chart, how to convert indicated speed to true airspeed, how to solve a triangle of velocities using the Dalton computer. Things like that.

The Dalton computer was an ingenious device, a bit like a slide rule, and is still in use by private aviators today. If you knew the wind speed you could read off the heading you needed to fly to make good your desired track over the ground. Conversely, if you knew the heading you had flown and the track you had made good you could determine the prevailing wind speed and direction.

In the long-range part of the course we got down to the real thing – maintaining an air plot, and learning how to fix position at night from the stars.

In initial training we flew in Expeditors, affectionately known as Exploders. These were small twin-engine planes, basically Beechcraft Model 18 aircraft modified for the military role. There were two navigator positions set up in the narrow, cramped cabin, each with a plotting table and communications. Later on in the long-range part of the course, we graduated to the much larger C47 aircraft, the military version of the well-known Dakota aircraft. This had, amongst other things, an astrodome for taking star shots as will be explained later.

Once in the Expeditor, I looked up from concentrating on my calculations to find to my amazement that we were flying only a few feet above the ground. The pilot had decided for a lark to 'beat up' his family home.

Actually quite a lot of the training was done on the ground. We were taught Morse code, meteorology, use of radar, and star recognition. And then there were the dreaded 'control plots'. We sat in a classroom with our charts in front of us and had to simulate a navigation exercise. Every now and then the supervisor would hand us a slip of paper, with some fresh information on it, perhaps a new fix or a new bearing. We had to assimilate this information and revise our plans accordingly. This required intense mental concentration, working against the clock. No hand calculators then; everything had to be worked out by mental arithmetic.

By far the most interesting part of the course was the astronavigation. The method was to take three stars in three different directions and measure their heights above the horizon. This then gave three position lines and a 'cocked hat' triangle where the lines intersected. The centre of gravity of the cocked hat was your most probable position. This procedure needed considerable skill to pull off. When I started, my position lines were way off, sometimes even off the edge of the chart. But it was surprising how with practice one got better and, in the end, my cocked hats became tight little triangles.

To 'shoot' the stars, which included planets and the edge of the moon if appropriate, we used the bubble sextant.

34

The first thing you had to do was to create the bubble. Then you held the object you were shooting in the bubble for two minutes while the sextant averaged the reading. The cabin of the C47 was unpressurised and cold – remember we were doing this in the depth of the Canadian winter – and ice would form on the inside of the astrodome. But we had a rubber hose delivering a hot air bleed from the engines that we could use to melt the ice. A further problem was to find suitable gaps in the cloud cover so that you could actually see the stars.

When we had saved enough money, a friend and I bought a car, an old second-hand Hillman Minx. The day after, my friend, who was an experienced driver, drove us down to the centre of Winnipeg and got out.

"It's all yours now," he said, "I'm going shopping. See you back at Base."

So there I was, an inexperienced driver – I had only just passed my test – in an unfamiliar car in the centre of an unfamiliar city having to drive back to Base on the wrong side of the road through a complicated and confusing grid of one way streets. Somehow I made it without mishap.

The car was a dreadful old banger. Its suspension had gone, but on the open road it went well enough. Of course, like most British cars of that time, it wouldn't start in the Canadian winter. But in the summer months it enabled us to get out and see something of the country. In the two weeks leave we had in the interval between the two parts of the course, we did a road trip to Tennessee and back in the Hillman.

It was now time to return to the UK. We travelled by train to Montreal where we spent a few days and then on to Halifax and from there by sea to Liverpool. I had now completed over a year of my two-year national service.

What was the Air Ministry going to do with me for the remainder of my time? After some home leave in Devon, where my parents now lived, I received the news that I was being posted to Flying Training Command Headquarters near Leighton Buzzard in Bedfordshire.

Leighton Buzzard, Bedfordshire

At that time, the Headquarters unit had a small 'communications' unit. I was attached to this unit which was based at Cranfield airfield some ten miles from the HQ itself. There was not much for me to do and my main responsibility seemed to be to look after the *Notice the Airmen* (NOTAMs) binder and keep it up to date by inserting update pages as necessary.

The unit had an Avro Anson for ferrying the VIPs on their visits, and several other aircraft including a Percival Provost and a two-seat version of the DH Vampire jet – very new at the time. In the Anson I had to navigate using a GEE[1] set. I had no idea how to use it. It was not taught at Winnipeg. It turned out that GEE, developed during the war, was one of the principal navaids used by the RAF at that time. However, luckily my ignorance did not matter as all our trips were at low level over familiar countryside by daylight in good summer weather, so no navigation assistance was required; the pilot knew his way without my help!

I was taken up in the Provost to act as an observer to spot if there were any other aircraft in the vicinity when the pilot was practising blind instrument flying. Sometimes though he would practise aerobatics. And sometimes he would let me have a go. I found the barrel roll easy enough, but the loop more difficult. You had to get up enough speed to make it

1   GEE stands for the letter 'g'.

to the top. On my first few attempts I found myself with the nose pointing vertically up, about to stall. The pilot would save the day by executing a quick 'wing over'. Then there was the Vampire jet which was interesting. The Vampire was one of the first jet aircraft to be developed and saw service at the very end of WW2. It was a single-engine single-seat fighter. But the one we had was the trainer version with two side-by-side seats. You sat right on top of the jet engine. You could hear it rumbling away beneath you. I hate to think what would have happened if it had thrown a turbine blade!

My national service time was now coming to an end and I had to decide what to do next. I could have carried on in the RAF. There was the possibility of taking a short service commission. Had I fulfilled my dream of becoming a pilot I would probably have taken this option. But somehow the secondary role of navigator did not appeal. So I decided to leave and try my luck in 'civvy street'. In retrospect, this was probably the correct decision. A service career is short-lived and there would have been few opportunities thereafter in the civil aviation world for navigators as the role was becoming increasingly automated. But curiously enough, as will be seen, I later found myself involved in air traffic control where my experience and knowledge of navigation came in very useful.

I really had no idea what I wanted to do so I approached the Cambridge University Appointments Board and they suggested I try Rolls-Royce who had a graduate engineering apprenticeship scheme. This sounded like a good idea so I went to Derby for an interview. They showed me a model of a vertical take-off airliner with dozens of small jet engines in the wings, one of their future concepts. This looked interesting so I signed up for a two-year graduate apprenticeship.

Chapter Four
THE STANDBY AIRCRAFT

Derby

The apprenticeship at RR involved moving between the different departments and spending time in each. In the workshop I learnt about the various methods of shaping metal; in the drawing office I learnt about third angle projections and the correct way to sharpen a pencil.

I also learnt something about how the working classes lived. I lodged in a back-to-back terrace house; I worked eight-hour shifts with only a fortnight's annual leave; I clocked in and out, and I was paid a weekly wage in cash in a window envelope every Friday. As a wage earner you had little job security. The firm could sack you at a week's notice. 'The staff' on the other hand were paid monthly and were on three months' notice. This kind of class distinction was the norm in industry in those days.

After two years in Derby, I concluded that I was not really sufficiently interested in engineering, nor did I have the right background, to make a success of a career in Rolls-Royce. I cast about for something that would make more use of my mathematical knowledge. I had heard of something called operational research, and did some research. It seemed that this was to do with improving the efficiency of organisations using mathematical and statistical methods. In fact, operational research was developed during the Second World War as a means of improving the effectiveness of military tactics. I bought a book (*Linear Programming* by Saul Gass, 1958) which described one of the mathematical techniques used, an application of linear algebra, a subject with which I was familiar from my Cambridge days. In the back was a description of an airline application. This sounded fascinating and rather up my street. The airline was British Europe Airways (BEA), one of the nationalised airlines set up after the war. I discovered they had an Operational Research group and applied for a job. They took me on in their Schedule Planning department.

It was the early sixties and BEA was flourishing. Foreign holidays were just beginning to become affordable for the average family. There was not much competition in the short-haul market and traffic was growing at about twenty per cent a year. The Schedule Planning department, as its name suggests, was responsible for drawing up the flight schedules. Plans were made several years ahead and then progressively refined as time went on. A major concern was to make best use of the aircraft fleet. When they were sitting on the ground, aircraft were not earning. So the aim was to maximise aircraft utilisation. Aircraft were becoming larger, more sophisticated and ever more expensive to buy and insure. The De Havilland Comet IV was being introduced to take over from the cheaper Vickers Viscount turbo-prop airliner and they wanted to make the most of its earning power.

One of the factors affecting aircraft utilisation was the need to provide spare capacity as a safeguard against delays and disruption to the schedule. This took the form of a 'standby aircraft' that could be substituted for any aircraft that had broken down or any flight that was unable to depart on time for any other reason. The question was: how many standby aircraft should be provided? Too many would reduce fleet utilisation; too few would worsen punctuality. This sounded like a typical operational research problem

– balancing one factor against another. In the past the management had worked on a crude rule of thumb. They had decided that five per cent of the fleet should be held back and not allocated to services. Thus a fleet of twenty aircraft would typically have one planned as a standby.

Operational research offered a more sophisticated and nuanced approach. It suggested that the operation of the schedule could be simulated so that disruptions and their knock-on effect could be tracked through the network. The effectiveness of standby aircraft could then be judged by the degree to which it reduced these unwanted and costly consequences. Such an approach pointed to the need for a computer to carry out the extensive calculations involved. It so happened that BEA was in the process of acquiring a computer from EMI which was just down the road from the airline's Head Office at South Ruislip where we were based. The machine was the EMIDEC 1100, an early type, one of the first using transistors instead of valves, but with very limited capacity. The machine was being acquired to process ticket sales data, but could equally well be used for mathematical work. It was therefore decided to go ahead and design and implement a simulation model.

It was agreed that the main causes of disruption to the schedule were three-fold: firstly engineering delays caused by unexpected faults that had to be fixed, secondly bad weather and thirdly air traffic control delays. The airline had good statistics on these factors so we could use these in a Monte Carlo type of model that picked values at random from appropriate statistical distributions. One of the commercial programmers did the programming. Looking back, it is amazing that he managed to do it at all. The EMIDEC had only 4K bytes of storage!

The resulting model could be used for a variety of studies.

One of the first was a study where to locate the standby aircraft. At that time, BEA had a fleet of DC3 Dakota aircraft, dubbed Pionairs by the airline, that operated domestic services between London, the Midlands and the Channel Islands. Aircraft night-stopped at Gatwick, Manchester and Jersey. The standby could be placed at any of these places. When we ran the model, we found no significant difference between these alternatives. Punctuality was unaffected by the location of the standby. It could be placed wherever was most suitable. Another study showed that the schedule could be 'tightened' by inserting an extra flight into the schedule without worsening punctuality.

The model was up and running so I moved on to the Operational Research group, which was then part of the larger Management Services department which included Work Study and Systems. We were asked to look at the cost of running the airline's ground operations, in particular the cost of passenger check-in. Here was another area where operational research had something to offer.

Glasgow was selected for study. A group of us travelled to Glasgow airport at Renfrew to review the operation of the airport and collect data. We stayed for a week at the New Caledonian Hotel which was near the airport and had tartan carpets. We carefully measured the time that the check-in staff took to deal with each passenger. When we got back to base, we fed this data into our operational research models.

One of the questions asked by management was whether it was better to have flight-specific check-in or 'common' check-in. The theory says that, all things being equal, common check-in is better. It reduces waiting time. Or put another way, we could operate with fewer desks and achieve the same level of service. However, when we looked at our data we found that with common check-in it takes significantly longer per passenger. So there was a balance of advantage and the optimum arrangement was not obvious.

It was an interesting experience to stand behind the check-in desks and see the process from the staff point of view. Many of the travellers were frequent flyers and well known to the staff. They arrived with their bags and greeted the staff with a cheery 'hello again' and proceeded with friendly chit chat.

The result of our study was that Renfrew was operating correctly at the correct staff level and there was no scope for savings. Although this null result was somewhat disappointing from a professional point of view – it would have been nice to have been able to claim some savings – it was no doubt welcomed by the local management. From our point of view at least we could claim that we had provided them with a sound method for staff budgeting in future years as the traffic grew.

Computer programming

The EMIDEC computer belonged to the Commercial department and all the programming was done by them. The standby aircraft model was programmed by Roy Claydon. I made friends with him. I would go into the office with him at the weekend when the machine was free and help him do the runs. There were rows of lights on the console showing the state of the registers. It was fascinating to watch the lights blink as the simulation progressed, sometimes halting for a bit while a queue in the model cleared.

I thought I would like to learn something about programming and found that Brunel College, then in Acton, ran an evening course on Elliott Autocode. The first evening, the teacher ran through the instruction set and introduced us to the machine, an Elliott 803. He then said, "I'm not going to give you any more lectures. Next time bring in some problems you would like to solve and try them out for yourselves on the machine." We were thus dropped straight in at the deep end.

The 803 was a paper tape machine. We prepared our input data on a modified typewriter, then queued up with our little strips of paper tape, fed them into the machine's reader and waited for the printed output. I still remember my feeling of surprise and marvel when I finally got my little program to work. The laborious calculations that would have taken so long to do by hand – it was a multiple regression – were done in seconds. What was more, you could so easily alter the input data and rerun the program. I was hooked.

The travel concession

No account of my time at BEA would be complete without mention of the travel concession. After a year's service you could fly anywhere on the network at ten per cent of the published fare, provided there was a seat free. So you could pop down to Heathrow and go to Paris or Rome for the weekend. In those days, fewer people were travelling, security was non-existent and the aircraft were smaller and more fun.

One notable trip I remember was one to Egypt with two friends, Sid Foster and David Belcher. Sid had done his national service in the Middle East and boasted that he knew all about dealing with Worthy Oriental Gentlemen. But when he went to recover his bags at the airport poor Sid found his camera was missing. The Worthy Oriental Gentlemen had seen him coming!

We visited all the usual tourist sights. We climbed the Great Pyramid – you could in those days – and inspected the signatures of Napoleon's troops inscribed in the stone at the top. We visited the Valley of the Kings riding mule back. We entered the tombs and marvelled at the brightly coloured frescos, something else that is now forbidden. We did not realise how lucky we were. On the way back we stopped off at Rome for a few days where Sid had a girlfriend.

Chapter Five
THE DECCA NAVIGATOR

C-E-I-R

My friend David Belcher had a friend, Len Taylor, who worked for a small computer outfit called C-E-I-R. I got hold of their sales brochure and read about some of the work they were doing. They were working on the airline crew scheduling problem. This sounded interesting. Moreover, it appeared that the staff there did their own programming which was something I was looking for. So I applied and went up to the West End for an interview with Philip Hughes, who later went on to found Logica. During the interview, the famous Martin Beale popped his head around the door with a long flow chart in his hand. I was impressed. I was then invited to an interview with the Managing Director, Maurice Kendall, the well-known statistician. Kendall was the co-author of the standard textbook I had studied – *An Introduction to the Theory of Statistics* by Yule and Kendall. So we were on the same wavelength.

I was offered a job and accepted. I joined C-E-I-R at the beginning of 1964. The company had just moved from the Turriff Building on the M4 to offices in Newman Street in London's West End in order to be near the IBM data centre where most of the company's processing work was carried out.

C-E-I-R was the UK subsidiary of an American company called C-E-I-R Inc.,[2] a computer firm based in Arlington,

---

2 I believe the initials originally stood for Council for Economic and Industrial Research although the firm was always referred to by its initials only and pronounced 'Seer'.

Virginia. The American company had developed, among other things, a proprietary computer program for solving large linear programming problems. The business idea was that the UK subsidiary would sell LP services, using the proprietary software, to London-based companies such as British Petroleum. Although this was the initial impetus, the UK firm soon branched out to sell other computer services.

The company was split into two divisions, the Consultancy Division, which analysed client's problems and wrote the appropriate computer programs, and the Operations Division which ran the programs. I joined the Consultancy Division.

At that time, the UK company had about fifty staff – a mixed bunch of mathematicians, statisticians and computer programmers and operators. Curiously, several of the staff had computer-related names. There was Maurice *Card*, Bill *Key* and Alan *Macro*. There was also a smattering of Americans with exotic names like Toby Riley lll.

Programming

It was a card-based operation. There were no screens in those days. Programs were held on eighty column Hollerith punched cards. We carried our programs around from place to place in large card trays. Cards had several advantages over paper tape, the alternative favoured by much of the British industry at that time, with the notable exception of ICL which because of its accounting history was also card based. Each instruction had its own card so editing was easy. You simply amended the cards that were faulty and slid them back into the deck. There was no need to do any splicing. The only downside was that if you dropped a card deck, getting the cards back in order was a nightmare unless you had taken the precaution of numbering them in advance.

The company had a small machine of its own, an IBM 1401, which served as an offline machine for the (large in those days) IBM 7090 computer across the road at the IBM data centre where the serious processing work was done.

We wrote out our programs on coding sheets. The programs were then punched into cards by dedicated punch girls and separately verified. The operational staff then loaded the card decks onto magnetic tapes, took the tapes across the road and ran them on the 7090. Output tapes were brought back and printed. More often than not the printout said 'compilation failed', indicating that the run had failed and we had to correct the errors and resubmit the job.

This round-about process took several hours from end to end. We had to wait a long time to see whether our programs had worked. This made for slow progress. Indeed, looking back, it seems remarkable that we managed to develop any programs at all. There was one advantage though. Runs on the 7090 were scheduled at set times of the day. As a programmer, if you could get your input ready for the late morning run you were then free to go for lunch in the knowledge that there would be no output to look at before mid-afternoon. And we were on the edge of Soho. There were some very nice restaurants to explore during the break!

The Decca job

The first job that I was given was to write a computer simulation to study the benefits of the Decca Navigator system, an area navigation system for civil aviation, rather like the GEE system I had come across in my national service days. Decca wanted evidence to show that their system was superior to the alternative point-to-point system, based on VOR/DME[3] beacons, which was favoured by the

---

3 Very High Frequency Omnirange/Distance Measuring Equipment

Americans. A choice between two was about to be made by the international aviation community. It was an important decision with far-reaching consequences.

This job was clearly up my street with my background in navigation and simulation, and I looked forward to getting stuck in. The client was a small air traffic control consultancy, GPS,[4] based in Ealing, West London – convenient for me as I was living in Ealing at the time. Their client in turn was the Decca Navigator company.

We took an existing busy air route and a typical day's air traffic. We assumed some aircraft were equipped with Decca and some not. We then constructed a second route parallel to, but offset from, the main route, and available only to aircraft equipped with Decca. We then ran a simulation showing how the traffic would behave under different conditions, and in particular how the Decca-equipped aircraft would behave. It seemed obvious that the Decca aircraft would benefit because they had a choice of routes to take. But by how much would they benefit? The answer was not obvious. The simulation was designed to answer this question.

We took Amber One (North), the airway out of Heathrow that passed over Daventry to the North, as the example route. Aircraft departure times were generated according to time of day to model peaks in demand. The simulation then mimicked the complex decisions taken by air traffic controllers when dispatching aircraft in real life from Heathrow.

I decided to program the model in CSL,[5] a high-level computer language that had been developed by John Laski and John Buxton at ESSO specially for simulation work. The

---

4 General Precision Systems

5 Control and Simulation Language

CSL language had certain useful features. It had a built-in model time advance mechanism, a pseudo-random generator and a set concept for describing queues. These features made programming simpler and speedier. But as with everything to do with computers, there was a trade-off; you don't get something for nothing in computing. The object code was complex and the model ran slowly. So although the CSL program could be written quickly, you paid for this in terms of execution speed. Our air traffic control model took about twenty minutes on the 7090 to simulate twenty-four hours of traffic. And time on the 7090 was expensive. An hour on the machine cost £250 which was a lot of money in those days.

I will always remember the first time we ran a 'production run' of the model. Paul Holden, our client at GPS, came up to our offices in Newman Street to supervise the run and select the input data. There was one data field that specified the proportion of aircraft equipped with the Navigator. The run was to take place in the evening. Paul booked a room in a nearby hotel so that he could stay as late as necessary. We took the tape across to the mainframe and set the model running. I kept my fingers crossed as the machine churned away writing its output to magnetic tape. I just hoped the run would complete successfully. I was greatly relieved when the run was over. It had worked and no bugs had appeared. We took the output tape back to the 1401 and printed out the results. We unfolded the fanfold output paper, eager to see the figures. But we could not see any results for the Decca-equipped aircraft to see how well they had fared, which was the whole point of the exercise. What had happened? Paul had a think.

"Oh no!" he cried. "I set the proportion of equipped aircraft to zero by mistake. There were no Decca-equipped aircraft in the run. The run is useless and I've wasted all that money!"

The next day, Paul's boss, Peter Haynes, came up to our offices to discuss the situation. Paul offered to pay for the run out of his own pocket. But Haynes brushed this offer aside. Looking around the conference table, he said, "This run was an essential part of the development process, chargeable to Decca, and if anyone says different, I'll kick their teeth in." Haynes was a businessman.

Of course, we then went on to run the model with the correct input settings. The results showed a dramatic reduction in departure delays for the equipped aircraft, which is what we were looking for. Haynes and Holden then went off to Decca to present the results.

We had given Decca some good ammunition for their case. But sadly for Decca, it was not to be. The international aviation community decided, instead of the Decca system, to adopt the alternative American point-to-point system, probably as a result of intense lobbying by US industry. So a unique opportunity for British industry was lost. Nowadays, of course, the advent of satnav and other avionics improvements has made this old aviation industry dispute irrelevant. Aircraft can fly point to point or offset from the airway without recourse either to the VOR/DME beacon system or the Decca Area Navigation system.

I wrote a number of other simulation models to run on the 7090, including one to study job shop scheduling at the finishing department of Stewarts and Lloyds steel works at Corby, Northamptonshire. I also published a small book, *Computer Simulation Models*, published by Charles Griffin, 1968, now out of date and out of print.

The Univac computer

In 1966, BP bought the UK part of C-E-I-R. The company changed its name to Scicon (short for Scientific Control

Systems) and we started using BP's own computer at its offices in the City – a Univac 1108 – instead of the IBM machine in Newman Street.

The UNIVAC 1108 came with something called a FASTRAND which was a massive rotating magnetic drum with a storage capacity of approximately 100MB. This was revolutionary at the times. It allowed us to store our programs 'online' on the drum for the first time. This was a great improvement. Now instead of having to carry around large metal card trays we only needed to keep small decks of edit cards. The programs themselves were always available on FASTRAND when we wanted to run them.

The first job that I did on the UNIVAC was a job for Seabridge, a shipping company. They owned a small fleet of about a dozen large ocean-going bulk carriers that traded across the Pacific from ports in the East Indies and Australia to ports on the west coast of America and Japan. They had heard, probably from BP, that computers might be able to improve the efficiency of their operations. We were contacted and we visited their offices in Leadenhall Street in the City, near the Baltic Exchange, where they explained the routes they used and cargoes that they carried.

The first problem we looked at was that of minimising the cost of backhauls. Their ships were basically carrying goods in one direction only; there was little or no cargo for them on their return journeys. And these empty backhauls were costing a lot of money. We realised that this cost could be minimised using linear programming – more specifically the well-known transportation variant of linear programming – the technique that got me interested in operational research in the first place. I spent time in their offices collecting data and formulating the computer model. I had to make some simplifying assumptions such as that the cost of a backhaul was proportional to the distance involved, something that would not necessarily be true in real life. I then ran the model on the UNIVAC using not the American C-E-I-R Inc. linear programming code designed for the IBM machine mentioned earlier but one developed in-house by us to run on UNIVAC machines. I think our client was quite impressed with the results. We had shown that we had appreciated an important problem right from the start.

We went on to look at how to optimise the schedule of outbound sailings, taking into account timing and other factors affecting operations, such as the availability of cargoes. I generated a set of feasible voyages for each ship in the fleet and then ran the LP program to select the set of voyages that would minimise the overall cost of delivering all the required cargoes. On the first run, the machine came up with a remarkably appropriate figure for the minimum cost solution – '1108', i.e. the designation of the UNIVAC computer that had performed the calculations!

I heard later that this scheduling system continued to be used after I had left. It was taken over by a group at AERE Harwell. They refined the model (using 'shadow costs') and continued running the system on a routine basis for many years.

Chapter Six
THE THIRD LONDON AIRPORT

At this time I was dissatisfied at Scicon and was looking for another job. I saw an advertisement for a position with the Roskill Commission which had just been set up to advise the government on the siting of a third London airport. The advertisement said they were planning to use cost-benefit analysis and this appealed to me. I felt that this was a much better way to make controversial public decisions than through the normal political process, although I came to change this view later, as I shall recount. The job also appealed to me because it involved aviation. I went for an interview at their offices at Templar House in High Holborn in the West End and accepted a post as a statistician in their Research Group.

A little bit of history is in order here. In the sixties, air traffic in the southeast was growing apace and the existing London airports of Heathrow and Gatwick were becoming full. It was therefore planned to expand the airport at Stansted in Essex to cater for the increase in traffic and to relieve congestion at the other two airports. But such was the opposition at the local planning hearing that the government decided to back off and instead put the whole thorny issue to a Royal Commission. A distinguished High Court judge, Lord Eustace Roskill JP, was appointed to lead the Commission. By appointing a High Court judge and employing cost-benefit analysis, the government hoped to make the final choice acceptable to the public. But in the event, this strategy did not work out as planned as we shall see.

The proposal to set up the Commission was announced in the Commons on 20th May 1968, by the then minister responsible, Anthony Crosland, see Hansard Vol 765.

It is generally agreed by all those who have expressed views on this matter that whatever plans one might have for other airports there is almost certainly a need for a third to serve the London area.

The terms of reference for the Commission were as follows.

To inquire into the timing of the need for a four-runway airport to cater for the growth of traffic at existing airports serving the London area, to consider the various alternative sites, and to recommend which site should be selected.

Why they specified four runways, I don't know, unless it was because the French had just stated constructing a four-runway airport at Roissy, just outside Paris, and they did not want to be outdone.

The Commission was set up in 1969. It consisted of seven distinguished members, assisted by a large research group of economists, statisticians, civil engineers and town planners, some specially recruited from the civil service and some like me from outside. Curiously, none, apart from myself, had a background in aviation.

A long list of about eighty possible sites was created. By the time I had joined, this had been whittled down to a short list of four. The idea was to measure costs and benefits at each site and then choose the one yielding best ratio of benefits to costs. In practice, this came down to estimating costs at each site.

The short list consisted of three inland sites and one coastal site. The inland sites, all north of London, were *Cublington*, near Wing in Buckinghamshire, *Nuthampstead,* in Hertfordshire and *Thurleigh* in Bedfordshire. The coastal site was *Foulness* on the Essex coast on an existing army firing range. Foulness was included as an alternative to an inland site and to test the popular claim that new airports should be built away from centres of population to avoid noise and pollution nuisance. For political reasons, Stansted was not included.

The Air Traffic Control problem

My job in the Research Group was to look at aviation aspects. The Air Traffic Control Authorities were concerned that the existence of a new four-runway airport close to Heathrow and Luton airports would cause problems in the airspace. Indeed, they considered that one of the four sites, Cublington, would be impossible to operate because of its close proximity to Heathrow. Their preference was Foulness because it was the furthest from existing busy air routes. I felt it was necessary to verify their opinions and if possible quantify the differences between the sites. I knew that my old friends at GPS had developed a computer simulation model of the London terminal area and this seemed a suitable tool for studying the problem. After some negotiation the Commission awarded GPS a contract to develop and run the model for us.

The first thing we had to do was to validate the model by running it against known present-day traffic arrangements to see how well it represented reality. It was well that we did this because later there was criticism of the model on the grounds that it failed to account for the human decision-making element in the system. But by that time we had obtained agreement that the model was sufficiently accurate for the purpose.

The ATC Authorities designed a route structure for each site and GPS ran a representative sample of traffic through their model. The results showed that there was no significant difference in terms of air traffic complexity between the sites. This was fortunate. It saved us from having to measure any difference in terms of cost, which would have been difficult. It also meant that Cublington was saved and could remain on the short list in spite of the objections from the ATC Authorities.

Radio Astronomy

One of the places it was said might be affected by the new airport was the Mullard Radio Astronomy Observatory near Cambridge. The very faint radio signals that they were receiving from distant stars might be distorted or even lost due to interference from aircraft going to and from the airport. With a colleague I visited the site. We were received and shown around by the head of the laboratory, Sir Martin Ryle, the distinguished scientist who later became Astronomer Royal. He took us to inspect his telescope which was a mile-long array of radio dishes installed on an old disused railway line. He then took us inside into a dark room to show us blips on a cathode ray tube which he explained came from one of the most distant objects in the universe. One could not help being impressed.

We showed Sir Martin the new air routes being proposed. These covered a large part of East Anglia.

"This would be a disaster for us," said Sir Martin. "We would have aircraft flying directly over our aerials. We would have to move."

"How much would that cost?" I asked, explaining our cost-benefit methodology.

"You cannot measure such a thing in pounds, shillings and pence,". he replied. "It is not just the cost of the physical move. Think of the loss to radio astronomy during the move. Every day counts for us, you know."

Sir Martin was the only person we came across during the whole cost-benefit project who flatly rejected the cost-benefit methodology. Everyone else was prepared to go along with it.

Choice of site

When all the cost and benefit figures were added up it was found that the three inland sites were very similar. But the Foulness site was significantly worse in terms of cost. The winner was Cublington, the site that the air traffic controllers had said was too close to Heathrow to be viable. The dominating cost in the analysis was the cost of surface access by road and rail, based on traveling time and distance, and this explains why Foulness came out worst in the analysis due to its greater distance from the places where most travellers started from.

The Commission accepted the result, but one member, Colin Buchanan, disagreed and put in his own minority report recommending Foulness instead. He could not go along with the loss of such a beautiful piece of Buckinghamshire countryside.

Timing of the need

As well as the choice of location, there was a second part to the brief: the timing of the need. When should the new airport be opened? This was a question of balancing traffic forecasts against existing airport capacity to see when the existing airports could no longer cope.

To assess future traffic demands we commissioned a

statistical survey of current passengers to provide a baseline for traffic forecasts up to the turn of the century. We turned these passenger forecasts into aircraft movement forecasts, taking into account the likely growth in aircraft size over the period.

There was some argument within the group over these forecasts. They were based on a constant growth rate. But why should traffic grow at a constant rate? This implied logically that eventually everyone would be in the air at once! But the economists pointed out that soon people would be taking more than one holiday a year. They would be making multiple trips. They were right of course. This is just what has happened, although the traffic has not grown as fast as we expected.

To assess capacity, I used a runway queuing model. This proved that, if the traffic forecasts were right, both Heathrow and Gatwick airports would be full by the time the new airport could be ready. There was need to start work on the new airport right away. But then we received an input from a group of researchers at the Royal Radar Establishment (RRE), based at Malvern, Worcestershire. They had been researching the possibility of using computers to assist the approach controllers at Heathrow and had concluded that such a system (Computer Aided Approach Sequencing) would increase Heathrow's handling capacity by ten per cent. This meant that the opening date of the new airport could be safely deferred by a few years. An urgent decision was not needed. The pressure was off. There was time for a rethink.

Some of us in the research team began to ask whether perhaps there really was a need for the new airport at all. Could there not be a better way to deal with the problem than a brand-new four-runway airport on a virgin site with all the disruption that that would entail? Could not perhaps existing airports be expanded to handle the extra traffic? In particular, could not Gatwick be given a second runway, doubling its capacity? We anticipated that the ATC Authorities might object that this would cause problems for them in handling the traffic. So we ran the GPS model again and showed that with the right route structure, air traffic control would not be a problem. As far as we could judge, future traffic increases could be coped with without the need for a new airport.

This led us to think that the original terms of reference were too restrictive. The terms of reference should have been 'find the best way of expanding capacity in the southeast' rather than 'find a site for a four-runway airport'.

But for the Commission it was too late to change course. They had already invested a lot of time and money on the search for a site. Public enquiries were being organised at each of the short-listed sites. The Commission was not about to say that all this work was nugatory because there was no need for a new airport.

In the event, of course, we on the research team were proved right. There was no need for a new airport. The existing airports were able to cope. And it is interesting to note that in the subsequent debate on the Commission's final report in the House of Commons on 4th March 1971, Anthony Crosland, now in opposition, criticised the Commission for interpreting their terms of reference too narrowly.

If there was any doubt that the terms of reference were flexible, I should have thought that it was finally dissipated by an answer which I gave on the same day to the then hon. Member for Sheffield, Heeley, Mr. Frank Hooley [...].

No, Sir, they [i.e. the terms of reference] do not entirely exclude that. It would be open to the Commission to say that there was no need for a third airport, or that it should be indefinitely postponed.

The end of the story

The Commission's final report recommending Cublington was published in January 1971. But in June 1970, six months previously, there had been a general election. A Conservative administration lead by Edward Heath had been elected replacing the labour government of Harold Wilson. The new government was not committed to the result and, perhaps influenced by the Home Counties' vote, overturned the decision and nominated Foulness instead – renamed Maplin Sands to made it acceptable to the public. A bill, the Maplin Development Act, was passed in Parliament and preliminary work was begun. But in July 1974 the Maplin plan was abandoned. In December 1979 it was quietly decided that the expansion of Stansted airport was the best solution after all.

Stansted? The place that had caused all the trouble in the first place? The wheel had come full circle!

Chapter Seven
THE NATIONAL AIRSPACE SYSTEM

I was now without a job. There were a couple of possibilities in the public sector, but they did not appeal. So I moved back to the private sector. I found myself a job at the Plessey Company. At that time, Plessey was implementing the Linesman Air Defence system and I worked on this for a time. My computer experience and general aviation background came in useful here.

I was then offered a position working on a new civil air traffic control system, then in the process of being installed at the London Air Traffic Control Centre (LATCC) at West Drayton just north of Heathrow. The system, called NAS,[6] was being acquired from the American FAA.[7] Its job was to track all flights operating in controlled airspace and produce reports. I was asked to lead a small team of software and system engineers from Plessey to assist with the installation and ongoing support of the system.

In the US, the FAA had already successfully installed NAS in each of their twenty enroute centres across the continent and the original idea was that the UK computer system would simply be the twenty-first example. Control of system development would continue to be by the US. But this changed. The UK authorities wished to control the system themselves. I was tasked with setting up a group to enable this to happen. But some of the Americans were reluctant to let go.

6　National Airspace System.

7　Federal Aviation Administration

The support job had two aspects. The first was the correction of faults arising during the running of the system. The second was the introduction of new functionality as required. We started with a group of six, which later grew to over twenty.

The NAS computer system itself was a complex beast. It consisted of a room full of IBM equipment – processors, printers, tape units, card readers, etc. put together specially for the task, designated the IBM9020D. The software was extensive and consisted of an online operating system – the Monitor – and a large suite of application programs.

Delivery and set-up

The hardware was delivered from the US by air freight. The equipment made it safely to Heathrow and was then loaded onto a lorry to go to the LATCC site. But enroute some of the equipment fell off and was damaged. Unfortunately, this was the only part of the journey that was not covered by insurance! So NATS had to replace the damaged bits at their own expense. A somewhat inauspicious start to the project!

A specialist IBM team flew to install the hardware. They knew exactly what they were doing and were super-efficient at it. They agreed just a week to do the job. They arrived from the US on a Monday and started right away installing the under-floor wiring. By Friday they had finished and were on their way back to the US. The CAA staff were amazed. Having spent several years with the British industry trying (and failing) to get a system to work, they found it hard to adjust to the idea of having a system delivered in just a week and what's more, one that was in fully working order.

The training course

The first thing we had to do was to get familiar with the

system. The FAA provided a comprehensive training course and all the documentation needed. The course was given by IBM. It was the course given to the maintenance staff at each of the US centres. The course was held at the Congregational Memorial Hall in the City and lasted ten weeks. The first week was devoted to an introduction to the system. It then went on to cover functionality, including input of flight data, storage, output of flight progress strips, recovery and fallback, adaptation data, etc. It also covered the software structure, the programming languages used, i.e. BAL[8] and JOVIAL,[9] and the system documentation. We discussed arcane subjects such as the difference between re-entrant and re-enterable code and the difference between switchover and startover, both of which NAS could do, according to circumstance.

At the end of the course we all repaired to Mother Bunches, a nearby pub, and had a merry time well into the night. Very few of us made it into work the next day.

NATS, our client, found us offices in Block A, a two-storey temporary building on the north side of the LATCC car park. Our first job was to get a version of the NAS software up and running on the off-line machine, which required some initial modification work. I well remember the first time it ran. Our whole team trooped across the car park to the main LATCC building where the off-line machine was situated. We were all eager to see what would happen when we tried to get the system to run. We typed in the GO message on the input-output typewriter, a modified IBM golf ball typewriter. And lo and behold the system immediately responded by typing out the date and time. NAS was up and running. We had achieved our first objective.

8 Basic Assembler Language
9 Jules Own Version of the International Algebraic Language

We enlarged the group by adding a dedicated test team. We set up a bureau operation on the offline machine. We created a document library to handle the extensive documentation of the system. And we started working on the software problems that arose during operation of the system. At first we needed quite a bit of help from the IBM staff who remained on site after the handover. But gradually our confidence grew until we could handle all the problems ourselves without help from the Americans.

User Charges

The first job we tackled was actually not to do with NAS itself but with user charges, the charges levied on airlines for using the ATC service. It happened that these were dealt with by a user charges office in Brussels. Details of UK-handled flights were sent to Brussels each month and the Brussels office then sent out bills to the airlines concerned and remitted the proceeds back to the CAA in London. This was an important job because this was how our client's operation got paid for and indirectly how we got paid!

At the time, the sending of data to Brussels was being done manually. A Ferranti computer printed out the flight details which were then reviewed and amended as necessary and sent on to Brussels over a datalink. We saw the potential of connecting NAS directly to the datalink and avoiding the manual process. Accurate data was now available for the first time via NAS and did not need a human review before being sent on.

The Ferranti computer, whose main function was to print flight strips, was now redundant and dispensed with.

There were other jobs too. We collected adaptation data for the radar processing function. We also replaced the operating system on the off-line machine. FAA had supplied

Royal Navy motor launch

Bomb map, Hampstead Garden Suburb

Hampstead Garden Suburb Archives Trust

HAMPSTEAD GARDEN SUBURB

HGS Bomb Map
Legend
- ⊚ V2 rocket
- ⤳ V1 flying bomb
- ◆ parachute land-mine.
- ● high explosive bomb
- · anti-aircraft shells
- ⊚ incendiaries c2000

Our house

Anti-aircraft shrapnel. (Scale inches)

On the river. I am furthest from camera

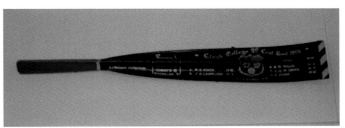

My blade

Beechcraft Explorer

Instruction number	Instruction field	card number

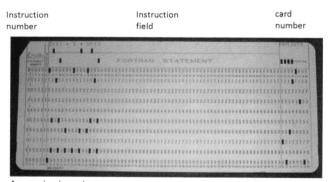

A punched card

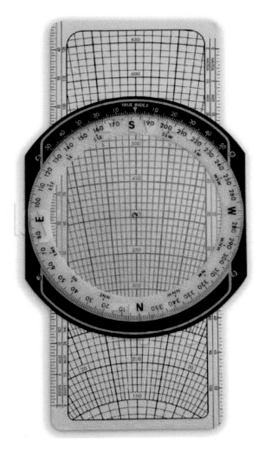

Dalton Computer

IBM 9020D Installation at London ATC Centre, 1975

Nimrod AEW aircraft in its hangar

us with NOSS,[10] a system written specially. We replaced it with MVS[11] which was the latest offering from IBM and provided a more efficient environment for bureau operations.

The Support operation was now up and running. CAA wanted to take it over from Plessey and run it themselves. My time at West Drayton was over. But funnily enough there was a sequel. Nearly two decades later I was called back to write a safety case for NAS as I shall describe in a later chapter.

Postscript

NAS went into operation at LATCC in 1974 and stayed in operation as the UK's principal flight and radar processing system for many years thereafter. Indeed, I believe it is still in operation at the time of writing nearly fifty years later. It has been modified and extended, its hardware has been replaced, but the core software has remained essentially the same over all these years. And it has had a remarkably good safety record.

---

10 I don't remember what the initials stood for (NAS Operational Support System?)

11 Multiple Virtual Storage

Chapter Eight
AIR DEFENCE

Improved UK Air Defence Ground Environment

I now left civil aviation for a while and moved over to the field of air defence which employed similar technology but with different objectives.

The Ministry of Defence (MOD) were looking to upgrade the air defences of the country with a new data handling system. They turned to Plessey for help. Plessey had successfully implemented the Linesman air defence system so it was natural for the MOD to look to the company for the new system. I was asked to lead a small team to study the computer and data handling requirements for the new system.

Whereas Linesman had been designed as a star configuration, with data being collected at a central location and then distributed, the new system was to be designed as a network to exploit the increase in computing power then becoming available. A network has obvious advantages in terms of resilience in the face of enemy attack. We looked at different ways of controlling the network in the event of damage.

I visited some of the radar stations. I was surprised at one station to be shown around by an American Colonel. I knew we were allies but I had not realised the transatlantic relationship was *this* close!

Although we had defined the main features of the system and knew what was required, the implementation went to

the American Hughes Aircraft company. I believe the MOD
had intended this in the first place. They had insisted that
Hughes place three of their staff in the project definition
team so that they were well placed to make a winning bid
for the subsequent work. Of course, Hughes were a natural
choice. They were leaders in the air defence field and had
supplied systems to NATO countries in Europe. MOD
probably felt more comfortable giving the work to them.

Airborne Early Warning Aircraft

Staying with air defence, I now left Plessey and joined a
small software company called Systems Designers Limited
(SDL) based in Frimley, Surrey, to work on the AEW Nimrod
aircraft. We were asked to study software and data handling
aspects.

The idea of an AEW aircraft was that at altitude its
radars could see further and detect incoming hostile aircraft
and missiles sooner than ground-based radars.

The AEW Nimrod was being procured by the MOD as a
replacement for the out-of-date AEW Shackleton which was
a propeller-driven aircraft with a radar mounted under the
forward fuselage which could only look downwards, so not
very suitable for an aircraft that had to scan the horizon. The
replacement, based on the more modern Nimrod airframe,
was planned to have two radars, one mounted in the nose
looking forward and one in the tail looking backwards.
Between them the two radars could then sweep the whole
360 degrees.

Stan Price, a friend of mine, had friends at the Woodford
site, near Manchester, where the aircraft was being
constructed. He was able to arrange for us to see one of
the prototypes – the so called 'hack' aircraft. We travelled
to Woodford and were met by Stan's friend who proceeded

to show us around. The aircraft was in a hangar on the far side of the airfield. She was a weird looking thing with large radar bulges protruding from nose and tail. She did not look very airworthy but no doubt she could fly.

At the end of our study we put in our report and held a large meeting at a hotel in Malvern, Worcestershire, to present the results to our MOD clients. But we got no formal response for our efforts. And later we heard that the whole Nimrod project had been cancelled. Whether our study had anything to do with the cancellation we will never know.

Switzerland

I next worked on a computer model of the Swiss defence data network. It had some similarities to the UK defence network I had worked on earlier. Only we were not allowed to know what was connected to the network and what type of data it handled; this was classified information. We did some research of the literature, as a result of which we decided to construct a closed analytical model instead of a simulation model.

Some of the project meetings were held in our offices which were now in Camberley. Others were held at the Swiss MOD Headquarters in Bern, Switzerland. On one occasion I decided to take some of the team across to Bern by air for the meeting. The company had a light plane which was available to senior staff to use for company business so I decided to use that for the trip. I could justify using the plane by the savings we would make on hotel expenses. This way we could do the trip in one day without a night stop. I planned this also as morale booster for the team, although unfortunately not everyone could come as the plane had only six seats. But as it turned out it was not such a good idea after all.

The company pilot did not have the necessary 'route experience' for a trip to Bern, so he asked a senior BA captain to come as well. I remember it all so well. It was a glorious June day. We assembled at 6 am at Gate B on the north side of the RAE[12] Farnborough site and took off for the four-hour flight to Bern which seemed to take forever. In the afternoon, after a long meeting with our Swiss clients, we headed back to the little Bern airport. We breezed through departures, bypassing the normal formalities like rock stars, and boarded our waiting aeroplane. It was then that things started to go wrong. By now it was late afternoon. Farnborough's runways would be closed for the night by the time we got there. The only option was to divert to Heathrow.

At Heathrow our little plane had to fit into the landing pattern among all the large passenger jets. The BA captain took control and steered us in. But he was used to sitting in a jumbo cockpit twenty feet above the runway. He put the little plane down with quite a bump. Goodness knows what it did to the undercarriage. We then had to find our way back to Farnborough to collect our cars. Management were not too pleased with our little jaunt which turned out to have been quite an expensive one for them.

We eventually got our computer model to run successfully and delivered to the client who was pleased with it because, being an analytical model, it ran much faster than the in-house simulation model they were developing at Bern. In fact, the model results were returned more or less instantaneously. This allowed the customer to try many variations very quickly indeed.

---

12 Royal Aircraft Establishment

Chapter Nine
THE SAFETY CASE

Up to this point SDL (rebranded SD) had been a successful company growing from small beginnings to about five hundred staff. But then the management made a mistake. They decided to float the company on the London Stock Exchange. Investors demanded continued rapid growth to maintain the share price. The company expanded too fast and got into financial difficulties. Redundancies were announced. Unfortunately, I found myself on the list of those being 'let go'.

On my last day at the company, I was clearing my desk and feeling rather low when the phone rang. It was my old friend Ray Ayres. He was in the same situation as myself.

"What are you doing?" he asked.

"Nothing," I replied.

"Well I am setting up my own company, would you like to join me?"

Obviously, I said yes. And he found me a job, starting immediately, at the CAA again. I must say I was very pleased to have something to go to. I did not want to sit at home all day with nothing to do. Sure, I now had my company pension, but still in my late fifties I was not ready to retire.

I set up my own service company and, working with Ray, was able to make a good living as a freelancer.

After a short spell at CAA in London, I took a job in Brussels at Eurocontrol, the European Air Traffic Control organisation. I didn't really want to work away from home but I found that I could do a weekly commute so that I was always back home for the weekend. I got up early on a Monday morning, drove to Heathrow, parked in long term, and caught the 7 am flight to Brussels. This got me to the office, which was in central Brussels – Rue Belliard – by late morning in plenty of time for lunch which we always took in the restaurant in the basement of the Berlaymont building. The return on Friday was even easier. I left the office at 3.30 pm and was home by 7 pm in time for the evening meal.

The Eurocontrol job could have been interesting. They were planning a ground-air data link. But progress was slow. Everything had to be agreed internationally. After six months I returned home and found a job with my old friends at West Drayton.

I may not have achieved much in my short stay in Brussels but the experience did wonders for my French!

Safety Management

In my new job I was asked by the CAA to develop a safety case for the NAS system. I felt well able to undertake the task as NAS had not altered much since I had worked on the system twenty years before and I knew the internals. However, I did not know anything about safety so they sent me on a safety management course.

The CAA had decided to adopt a system-based approach to safety management, a relatively new approach developed as a result of the Piper Alpha disaster in the North Sea in 1988. Historically, safety had been a question of conforming to standards and rules and regulations – a prescriptive approach. But the Piper Alpha Enquiry had concluded that "the prescriptive approach was not adequate".

A new 'system based' approach should be adopted which considers the specific nature of the system in question and sets safety targets expressed as probabilities that must be shown to be met. The results of this analysis should be recorded in a safety case.

A safety case is 'a structured argument, supported by evidence, intended to justify that a system is acceptably safe for a specific application in a specific operating environment' (Ministry of Defence definition). The weasel words here are 'acceptably safe' as we shall see.

The Burgundy Book

Following this approach, the CAA devised its own procedure, recorded in a safety management manual, known as 'the Burgundy Book' due to its dark red cover. There were four parts corresponding to the four stages of the project life cycle: requirement, production, transition to operation and operation. The idea was to show that the requisite level of safety defined in Part One was achieved at each subsequent stage in the project.

For the NAS safety case, then, the logical place for me to start was the Part One, the requirement. However, I did not feel competent to write this on my own. I looked to the CAA for help.

It turned out that this was not an easy task. How do you define what is 'acceptably safe'? Or put another way, if safety is to be determined by the probability of not causing harm, how do you determine what is an appropriate probability level? The UK safety community had come up with the ALARP principle, which states that risk, expressed as a probability of causing harm, should be **A**s **L**ow **A**s **R**easonably **P**ractical. But of course, this only restates the problem without solving it. In the end it comes down to a judgement of what is 'acceptable'. The problem is that 'safe' in this context is a term that does not admit of a precise definition. This leaves open the possibility, in the case of an accident, of legal action being brought against the system supplier or operator by someone disagreeing that the system is safe. This is why the Americans have retained the prescriptive approach which protects the system supplier or operator if he can show he has obeyed the rules prescribed. But I digress.

The CAA tried to avoid the problem of writing the Part One document for NAS.

"Oh yes, I'll have a look at it on the train coming in tomorrow," would say one. But then on the morrow the remark would be, "I don't think I am the right person to sign this. Perhaps so and so would be more appropriate."

This passing the parcel went on for some time. In fact, we never did get a Part One for NAS.

I avoided the problem by approaching the problem from the other end. Instead of defining a requirement and proving that the system met it, I turned the problem around and first defined the hazards and then asked if these were acceptable.

The knowledge of this reluctance to sign off the Part One came in handy when later I was involved with Stan Price in defending the safety aspects of the design of the new Oceanic system, as I shall describe later.

The Hazard Analysis

In the hazard analysis there were two failure modes to consider: fault and loss, the first was when the displayed data was erroneous; the second was when displayed data was missing. For each type of data, these two modes had to be considered as a hazard.

I constructed a spreadsheet with a row for each identified hazard, showing the precautions taken to prevent its occurring, the mitigations available if it did occur and the consequent assessed risk level. This demonstrated that, even if not totally risk free, NAS did not pose a serious danger to life and limb. Neither failure mode was considered by the CAA to be dangerous – inconvenient yes, and perhaps costly, but not a threat to life and limb. The controller could always handle and recover the situation safely – even in the event of a total system failure. It was also true that such hazards were extremely unlikely to occur as plenty of safeguards had been built into the software in the first place.

The conclusion was that NAS was safe. No further measures or redesign was necessary.

Subsequent events have shown NAS indeed to be very reliable and safe: the ATC system as a whole has hardly failed at all in the last thirty years.

While working on this project I began to worry about my own position. What if there was an accident? What if the blame was then placed on an inadequate safety case? What if the prosecutors argued that the harmful event could and should have been foreseen when writing the case? I consulted a lawyer who assured me that I was fireproof because I was working under instructions and my safety case would be signed off by the CAA. In any case, a prosecutor would be unlikely to go after a small one-man company like mine with very limited funds.

The Safety Critical Systems Club

I was invited to join the Safety Critical Systems Club which had recently been formed. The club ran a series of conferences and seminars. I spoke at some describing my experiences at West Drayton.

There was also a series of two-day workshops run by Stan Price under the title 'What goes on out there?'. The workshops were designed to compare industrial practice with the activities of academia and learn lessons. Stan asked me to act as rapporteur. This involved producing a report of the proceedings which was then published. Stan had good taste in the hotels he chose for the workshops. They were always most comfortable and located in agreeable countryside. I well remember one workshop that was held at the Izaak Walton hotel in Dovedale. It was summertime, the weather was fine, and in the evening after the workshop we walked for miles along the beautiful River Dove discussing the topics of the day.

I believe the club is still going today, and has recently celebrated its thirtieth birthday.

The Oceanic ATC project

About this time, NATS was planning to replace the computer system at their ATC Centre at Prestwick in Scotland which handled flights over the Atlantic Ocean. They decided to procure the replacement by means of a Public Private Initiative (PPI), a method popular at the time in the public sector for funding infrastructure projects such as new hospitals. To NATS this seemed like a good wheeze. The supplier would take all the development risk and bear all the initial cost. Effectively, NATS would get the system for free. So far so good. However, there was an obvious snag. The system would belong to the supplier and NATS would then have to pay for using it. Over a thirty-year period, a typical system lifetime, this could amount to a hefty sum. And in addition, NATS would have to pay for any changes they wanted to make to the system.

Electronic Data Systems (EDS) was selected to do the job. In fact, EDS was one of the few software houses of sufficient size, experience and financial muscle that could undertake such a complex task at its own financial risk. EDS assembled a large team and began work. But after some time NATS became dissatisfied with the project and decided to cancel it and pursue a different approach. I don't know if this was because EDS were not sticking to the specification, as NATS alleged, or because NATS came to realise that they would eventually be charged for every change they wanted to make – and with an ATC system there are always many changes wanted. Not only would this be expensive but it might inhibit NATS from making necessary changes due to cost – which might even have safety implications.

Anyway, for whatever reason NATS went ahead, cancelled the project and terminated EDS's contract. The software team was now out of work and the independent contractors had to go. Alan Martin, one of the independent contractors and a friend of mine, told me how it happened.

"There I was working away at my desk as usual when I was called into the boss's office.

'Now Alan, I've got good news and bad news, which do you want to hear first?'

'Give me the bad news,' said Alan.

'The bad news is that the project's been cancelled and you are no longer required.'

'OK, so what is the good news?'

'You can go on holiday!'"

Some consolation!

The Legal Case

Naturally EDS wanted to get its money back and took NATS to court for breach of contract. EDS hired a firm of lawyers to conduct their case and the lawyers in turn wanted expert witnesses to provide technical evidence in support of their claim and to refute the counter arguments of NATS. In particular, the lawyers were worried that NATS would play 'the safety card'. They would argue that the EDS design was unsafe and could not be put into service.

To cover this eventuality, they asked Stan Price if he would be prepared to act as an expert witness as he had worked on safety issues as a member of the safety critical systems club and had worked on legal cases before. Stan in turn asked me if I could act as his 'researcher'. I was actually in an ideal position to help because I had worked for both organisations in the dispute and knew both sides. I had worked for NATS on their NAS safety case and knew how their safety procedures worked, and I knew EDS because they had taken over SDL, the company I used to work for and I knew many of the staff involved in the project. In fact, I knew people on both sides of the dispute!

One of NATS's complaints was that the system was inadequately redundant because it was based on a twin processors concept – a main and a standby and this might not give the reliability they needed. I was able to point out, from my own knowledge, that NATS's own system, NAS, had a similar design. They could hardly say that their own system was unsafe or unreliable! This neatly answered their complaint.

Stan and I then put our heads together. We went on the attack. It seemed to us that if NATS were to accuse EDS of coming up with an unsafe design they would have to base this statement on something. Rather than simply complaining, they would have to declare the level of safety they were looking for. Knowing the NATS safety system as I did, I reckoned that they would have had to declare their safety requirements in Part One of the safety case. But also remembering the problems they had with writing the Part One of the NAS safety case I was interested to see if they had managed to write one for this project.

"We should ask to see their Part One Safety Case," I said.

"We are entitled to see it under the 'Disclosure' rules that apply in legal cases."

So we told the lawyers to ask for it. But it seemed that NATS was having some difficulty locating a copy. Time went by. They asked again and still no Part One. Eventually, NATS had to admit that there was no such document.

So now NATS could not sensibly claim that EDS were not meeting the required safety standards. There were *no required safety standards*. Moreover, NATS were now in the embarrassing position of having to admit they were not following their own procedures. The Burgundy Book did have some weasel words allowing a project to avoid a Part One in certain circumstances, nevertheless the thrust of the four-part safety case was clear: the requirement should be specified in Part One.

The court had been sitting for several weeks when one day something unusual happened. Someone at the back of the court interrupted the proceedings, approached the judge's table and handed in a slip of paper. What was this? The judge read the figure on the paper, nodded and remarked, "This seems sensible." Then he stood up.

"The case has been settled out of court. The proceedings are now closed."

And with that he walked out of the courtroom.

But what was the figure on the slip of paper? We were never told. Rumour had it that the figure was £10m. Some say more. So EDS got their money back. But this was not much comfort to those who had lost their jobs. Some never worked again. And the UK had now lost the capability of implementing large-scale command and control systems of this type. NATS went on to acquire a system from Canada instead.

After 'the Case' I was thinking of retiring but I found myself one last job at Swanwick in Hampshire where NATS were building a brand-new ATC centre to replace the one at West Drayton. They wanted a statistician to analyse project data and measure progress.

The project had reached the testing stage. Every month they rebuilt the system to incorporate fixes to bugs that had been found. The rebuild inevitably created new bugs and these had to be corrected in turn and the solutions incorporated in a further rebuild. And so on ... My job was to see how close the project was to getting to a situation when all the bugs had been eliminated.

At every build I counted the number of bugs that had been found since the previous build and the number that had been addressed and 'corrected' in this build. I plotted these as two curves showing cumulative totals against time. I had a Find Curve showing the history of the discovery of errors and a Fix Curve showing the history of correction of errors. In principle, the Find Curve should start flattening out as the software becomes more reliable and the Fix Curve should start converging on the Find Curve showing that all finds had been fixed and there were no more errors left in the system. The only problem was that the two curves did not behave like this. They both continued to trend upwards and to run parallel to each other. This implied that the end of the project was not yet in sight. The project was destined to take a lot longer than the project team thought. I ran a ruler over the curves to show my boss that the curves were not converging.

"No need to show these results to management," he said. "We don't want to frighten the horses!"

But my statistics were right. It was another two years before the system was reliable enough to put into service.

At this point I decided that I deserved some leisure time to pursue other interests. I was sixty-five and concluded that it was time to hang up my hat and retire.

Chapter Ten
EPILOGUE

In my long life I have seen great change. When I started out, working conditions were quite primitive. Computers were just coming in. There were no screens – the bitmapped display had yet to be invented. You worked with paper and pencil. There was no internet. You communicated by handwritten notes and the telephone, if you were lucky enough to have one on your desk. Copiers were messy liquid affairs.

In everyday life too I have seen great changes. When I was growing up we did not have a car, a television or a telephone – let alone a mobile one. There was no internet, no central heating, no supermarkets. Foreign holidays were the preserve of the few. How did we manage? We got on. Public transport was efficient and cheap. The shops were within walking distance and if you wanted to go further there was always the bicycle – traffic was light and cycling easy. The radio and cinema entertained us and kept us informed. Letters were delivered twice a day and urgent messages could be sent by telegram. And we heated the house with coal and were thus independent of the large power utilities that rule the roost today.

I have seen great social change too. The world I grew up in was predominately run by men. The manual labour force was exclusively male, as were the professions. Women mostly stayed at home and looked after the family. *Kinder, Küche, Kirche* was the rule of the day. Homosexuality was a crime punishable by a prison sentence – how different from today's world where alternative lifestyles are celebrated.

Yes, what changes I have seen. But not everything has changed. The buses still run up Finchley Road and stop at Temple Fortune where I used to get off when coming home from school. The 'Torbay Limited' train still departs Paddington for Torbay every day, albeit without the name or the restaurant car. And above all the seaside does not change. Children play on the same rocks and investigate the same tiny marine creatures in the same rockpools at low tide as Christopher and I did all those years ago.

ACRONYMS

AA	Anti-Aircraft
AEW	Airborne early Warning
ATC	Air Traffic Control or Air Training Corps
AWACS	Airborne Warning and Control system
BAE	British Aerospace
BEA	British European Airways
CAA	Civil Aviation Authority
CAAS	Computer Aided Approach Sequencing
EDS	Electronic Data Systems
FAA	Federal Aviation Administration
FDPS	Flight Data Processing System
GPS	Global Positioning System
IUKADGE	Improved United Kingdom Air Defence Ground Environment
IT	Information technology
LP	Linear Programming
MOD	Ministry of Defence
NERC	New Enroute Centre
NATS	National Air Traffic Services
PPI	Public Private Initiative
RAF	Royal Air force
SDL	Systems Designers Ltd